RADES K-2

SCIENCE

A CLOSER LOOK

Technology
A Closer Look

Macmillan/McGraw-Hill

Content Consultants

Robert L. Kolenda
Science Coordinator, K-12
Neshaminy School District
Langhorne, PA

Paul Mulle
Science Supervisor
Camden City Schools
Camden, NJ

Patricia Cavanagh
Elementary School Teacher
Merrimac Elementary School
Holbrook, NY

Pat Marengo
Sixth Grade Teacher/Technology Specialist
District 58 Elementary District
Downers Grove, IL

Students with print disabilities may be eligible to obtain an accessible, audio version of the pupil edition of this textbook. Please call Recording for the Blind & Dyslexic at 1-800-221-4792 for complete information.

Internet Disclaimer
Visit www.macmillanmh.com to learn more about technology. You will also find links to other Web sites. These Web sites are not run by Macmillan/McGraw-Hill. When using the Internet, be safe and protect your privacy. Make sure a teacher, parent, or guardian is around. Never tell someone your full name, address, passwords, or other personal information. Do not respond to e-mails or messages from strangers.

The McGraw-Hill Companies

Send all inquiries to:
Macmillan/McGraw-Hill
8787 Orion Place
Columbus, OH 43240-4027

Printed in the United States of America

ISBN-13: 978-0-02-286123-0
ISBN-10: 0-02-286123-8

7 8 9 10 WEB 15 14 13

Contents

Lesson 1

We Use Tools

Vocabulary

tools

technology

X-ray

Tools help us do things. What tools do you use every day?

You can use a shovel to dig in soil.

You need a hammer to build a bird house.

Scissors are tools, too.

All the tools and ideas we use are called **technology**. Technology can make our lives better. Technology helps us travel, communicate, and stay healthy and safe.

People have always used tools and ideas. New ideas and tools can make work easier.

broom

 **What are tools?**

Tools Then and Now

Read a Photo

How are these tractors different? How are they the same?

Technology Can Help Us

Technology depends on science. Science depends on technology, too.

Tools help scientists experiment and explore.

A telescope is a tool. It lets us see faraway objects in space.

▼ **Technology can tell us about the weather.**

Sometimes helpful tools can harm us. We must be careful how we use them.

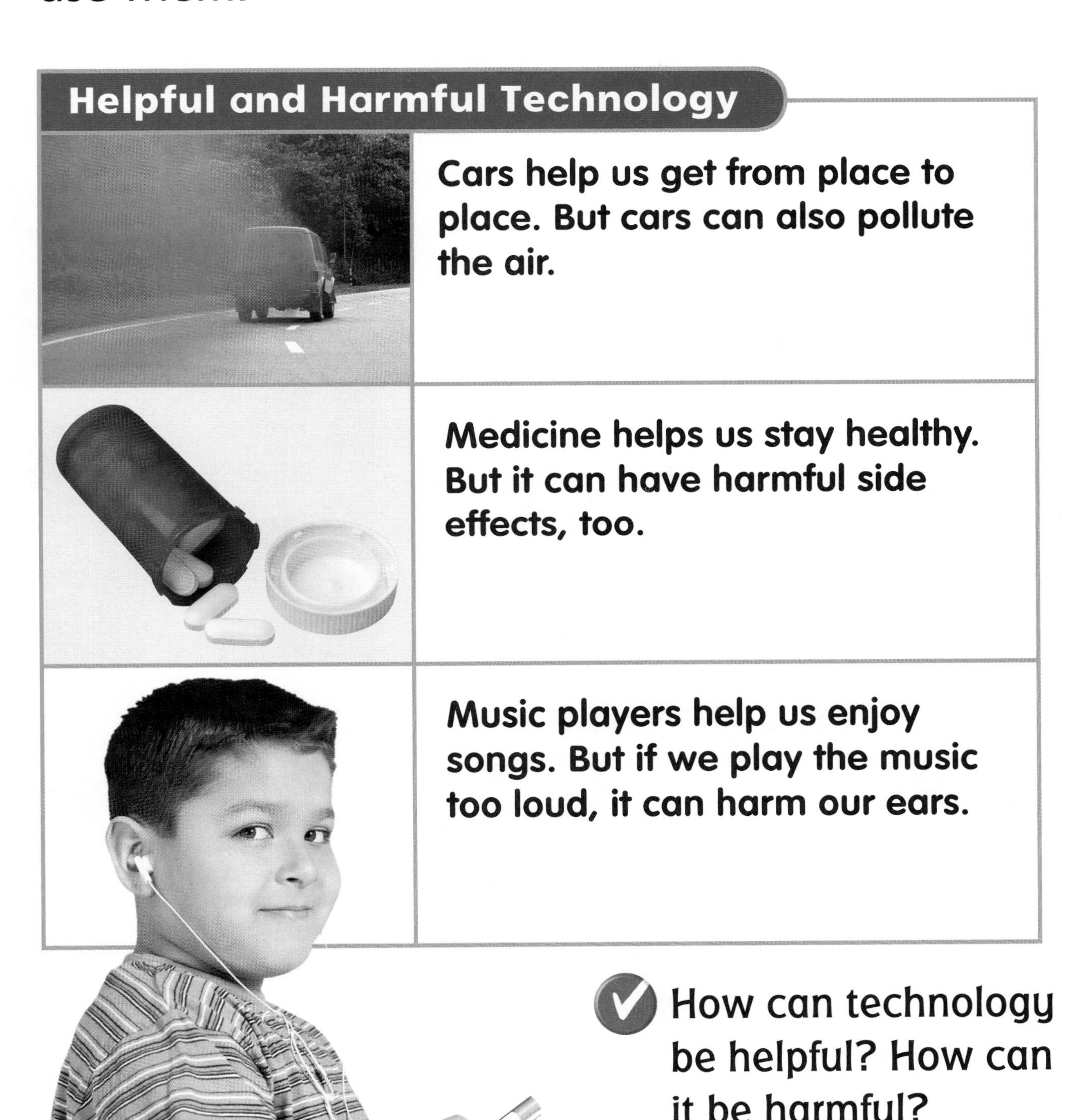

Helpful and Harmful Technology

	Cars help us get from place to place. But cars can also pollute the air.
	Medicine helps us stay healthy. But it can have harmful side effects, too.
	Music players help us enjoy songs. But if we play the music too loud, it can harm our ears.

How can technology be helpful? How can it be harmful?

A Tool to Look Inside

Can you see inside your body? No, but a machine can. It takes special pictures. They are called **X-rays**.

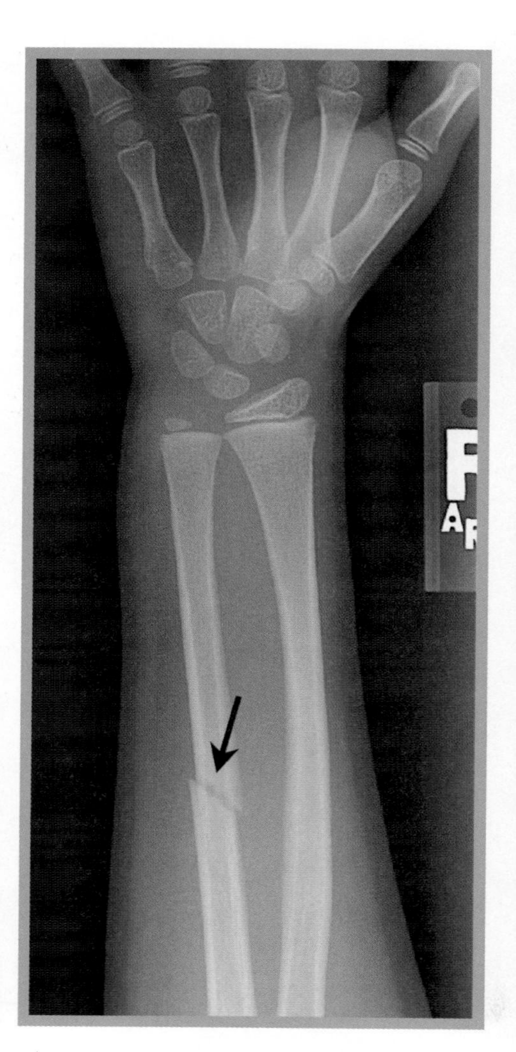

▲ This is an X-ray of a broken arm.

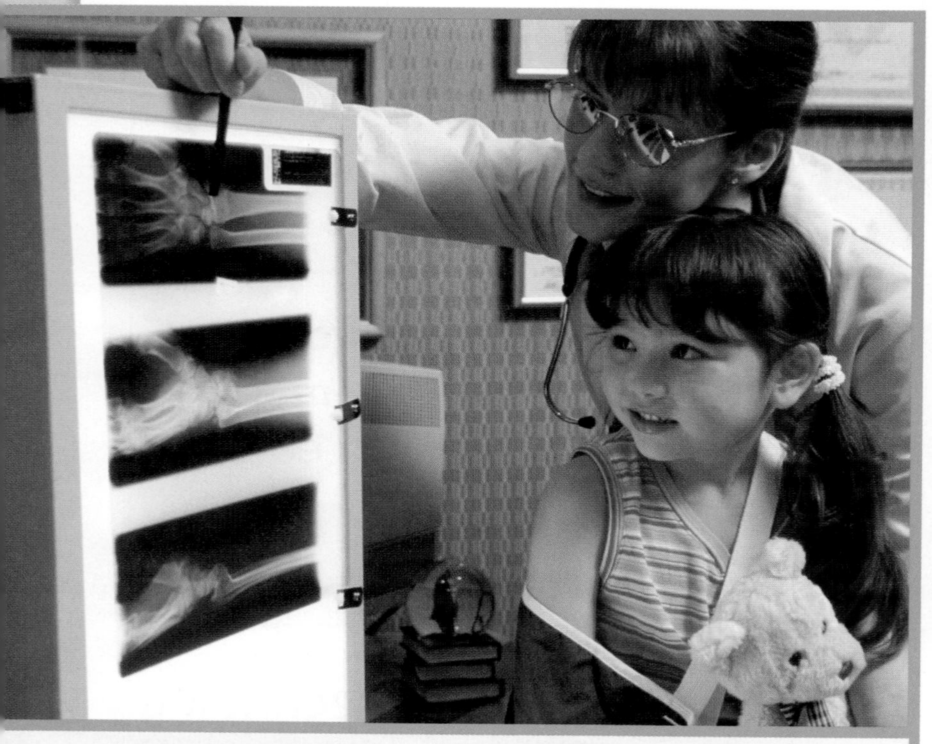

X-rays go through skin and muscle. They can not go through bones. So bones show up on X-ray pictures.

Dentists take X-rays. They want to see how your teeth are growing. They also look for cavities.

Doctors and dentists look at X-rays to see what is wrong inside. Then they can fix it.

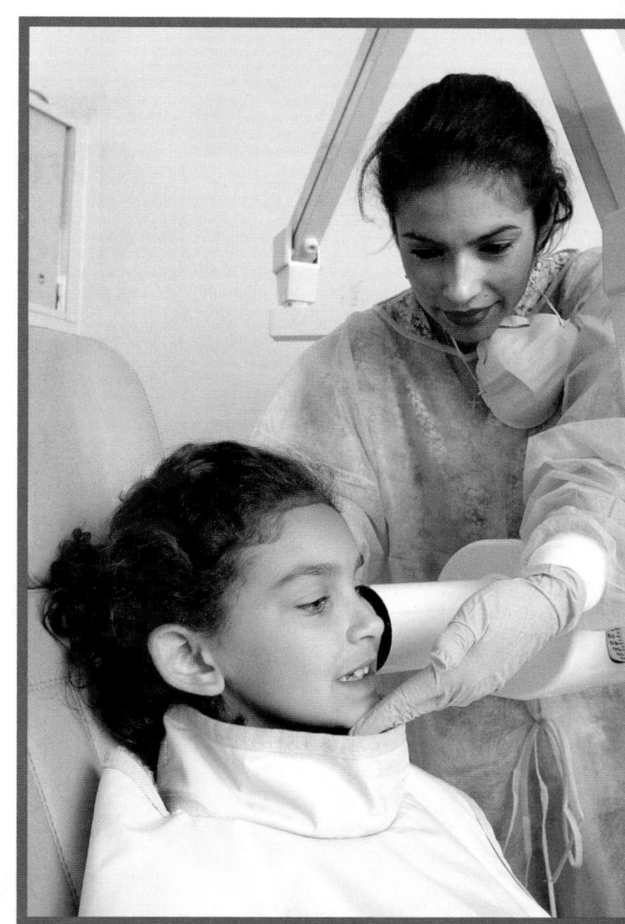

This girl is getting an X-ray of her teeth.

▼ **What do you notice about the teeth in this X-ray?**

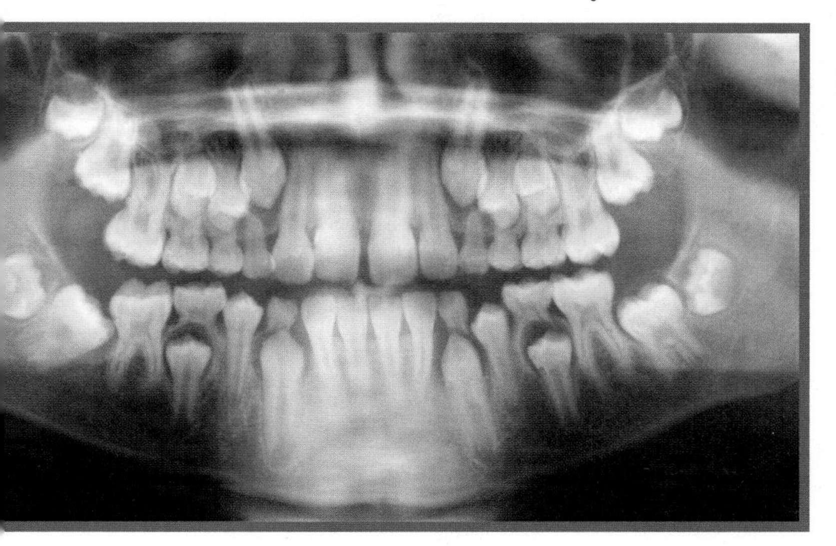

Talk About It

Summarize. When might a doctor or dentist need to see what is inside of you?

Tech Activity

You need

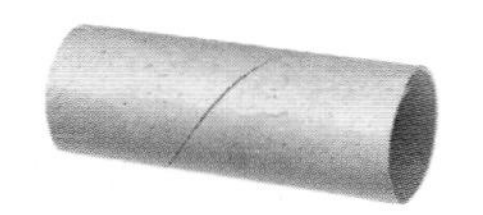

paper towel tube

scissors

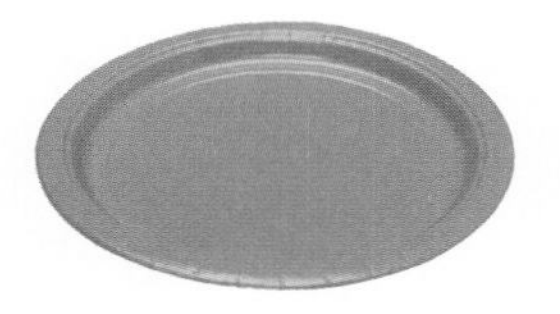

paper plate

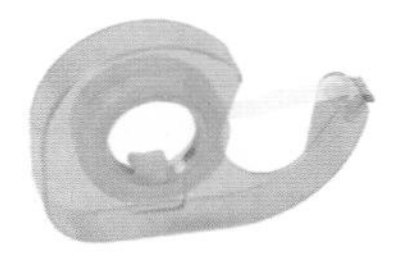

tape

glue

Make Your Own Tool

What to Do

1. Use a paper towel tube as the handle for a broom.
2. Cut paper. Put it on the end of the paper towel tube.
3. **Investigate.** Use your broom to sweep an area of your classroom. Collect the dirt in a paper plate.

Draw Conclusions

How is your tool like a real broom?

Think, Talk, and Write

Complete each sentence.

tool

X-ray

1. Something that helps us do things is a ________.

2. A special picture of inside your body is an ________.

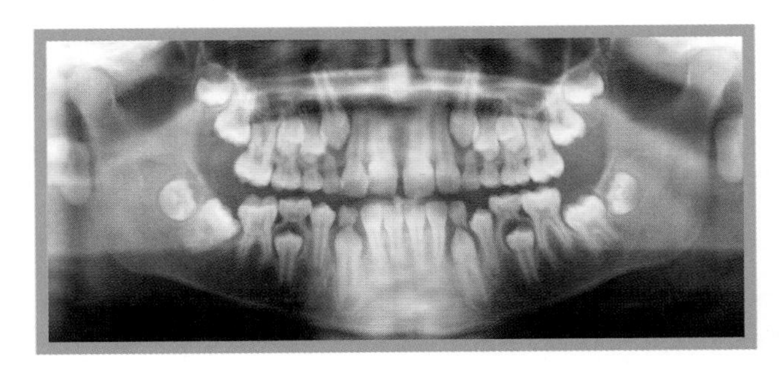

3. What are some ways technology helps us?

4. What are some ways technology harms us?

Art Link

Which tool do you use most to help you do work? Draw yourself using that tool.

LOG ON e-Tech Link For more on tools check out www.macmillanmh.com

Lesson 2

Materials and Their Uses

Vocabulary

natural resources
properties
mint

Materials are either made by people or nature. Materials made by nature are called **natural resources**. Natural resources come from Earth.

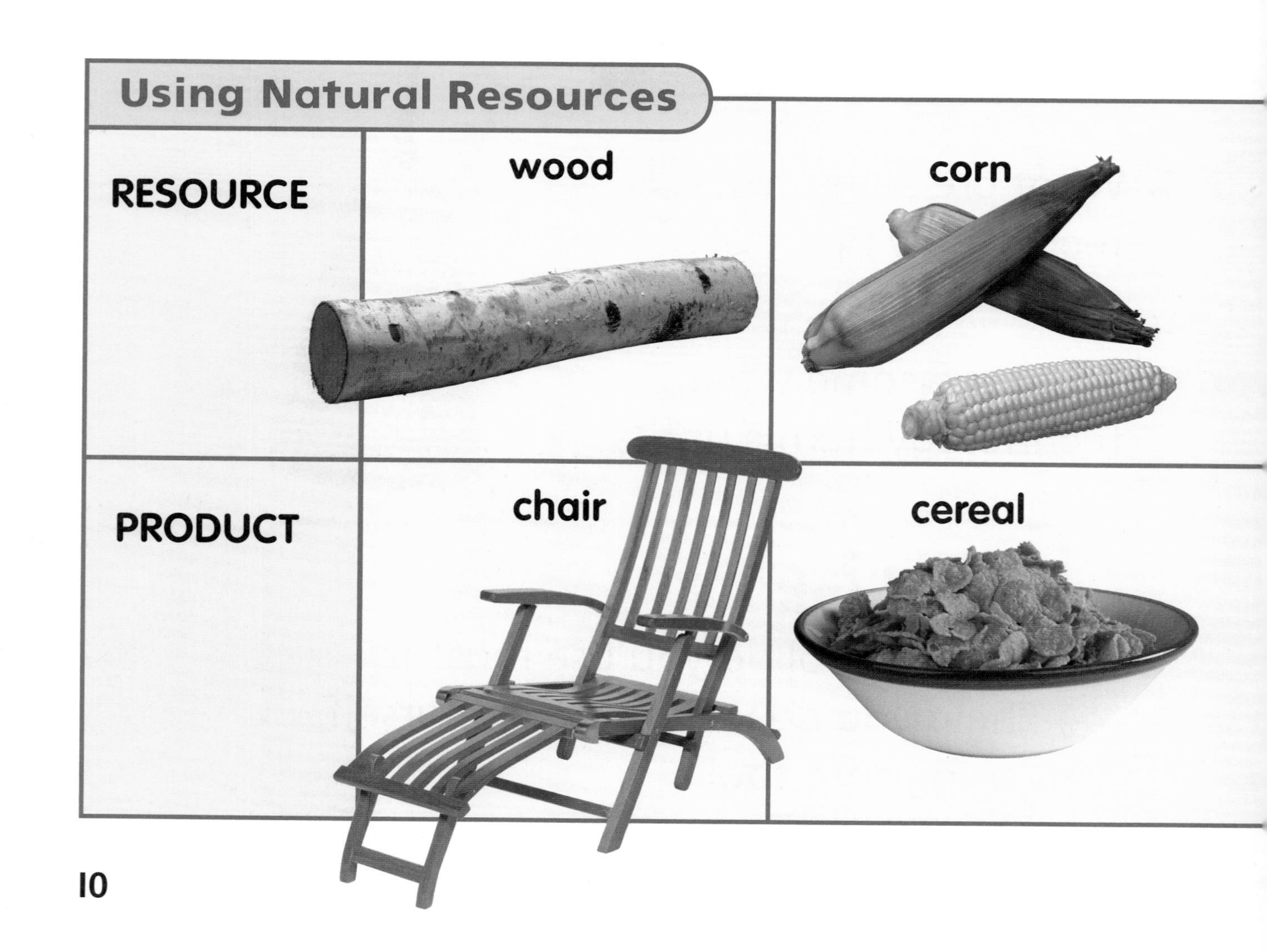

Plants and animals are living natural resources. Land, minerals, water, and air are nonliving natural resources. We use both kinds of natural resources.

✔ **What do you see in your classroom that is made from wood?**

Read a Chart

What can we use wood to make?

Properties of Natural Resources

We use natural resources to make new materials. Materials have different properties. **Properties** are how something looks, feels, smells, or tastes.

Properties make materials good for some things, but not for others.

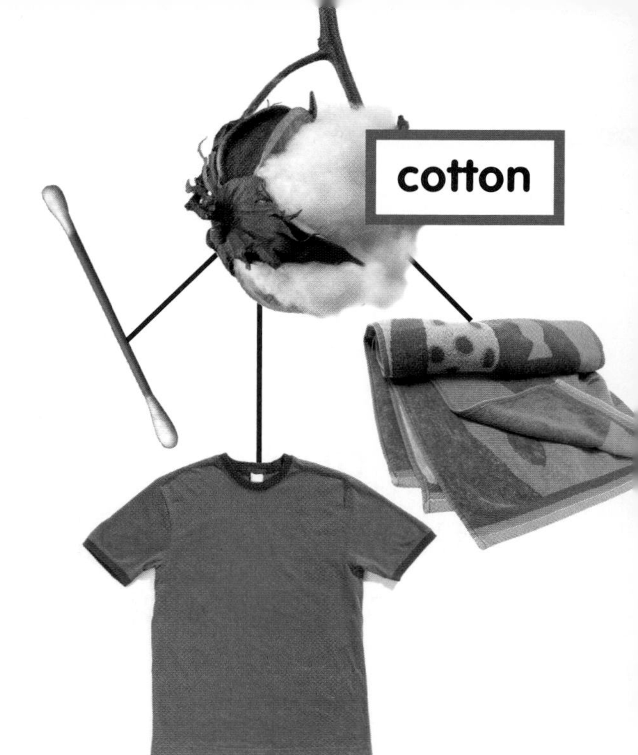

Cotton is a natural resource. Softness is a property of cotton.

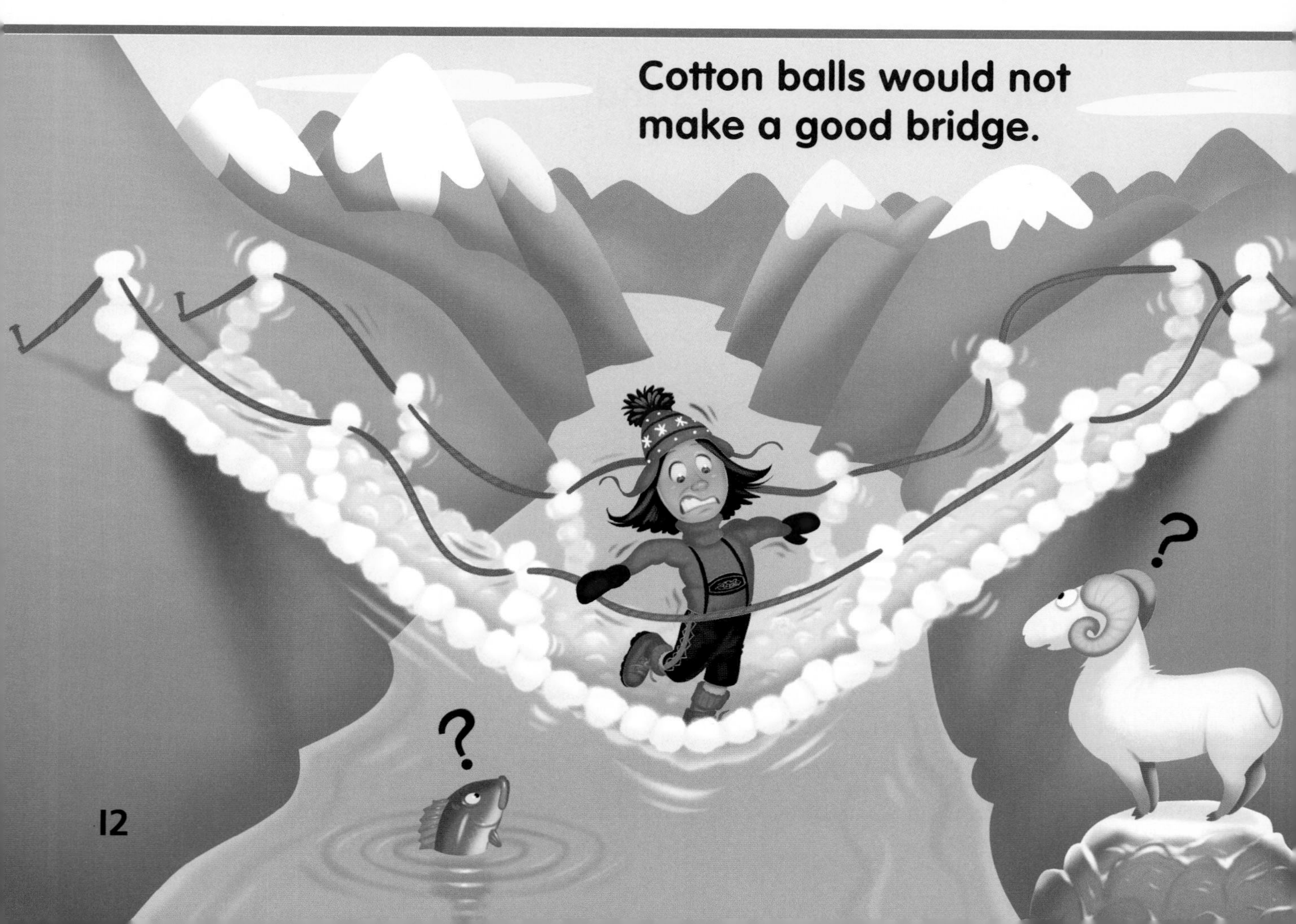

Cotton balls would not make a good bridge.

We use many things made of plastic. Plastic does not grow in nature. People make plastic. It can be soft or hard. It is very strong.

Some natural resources are limited. That means they will not last forever.

✔ How are cotton and plastic different?

You can help save Earth's resources by recycling things made from plastic.

Making Money

Do you know where money comes from? A **mint** is a place where money is made.

First, metal is melted. Then it is poured out.

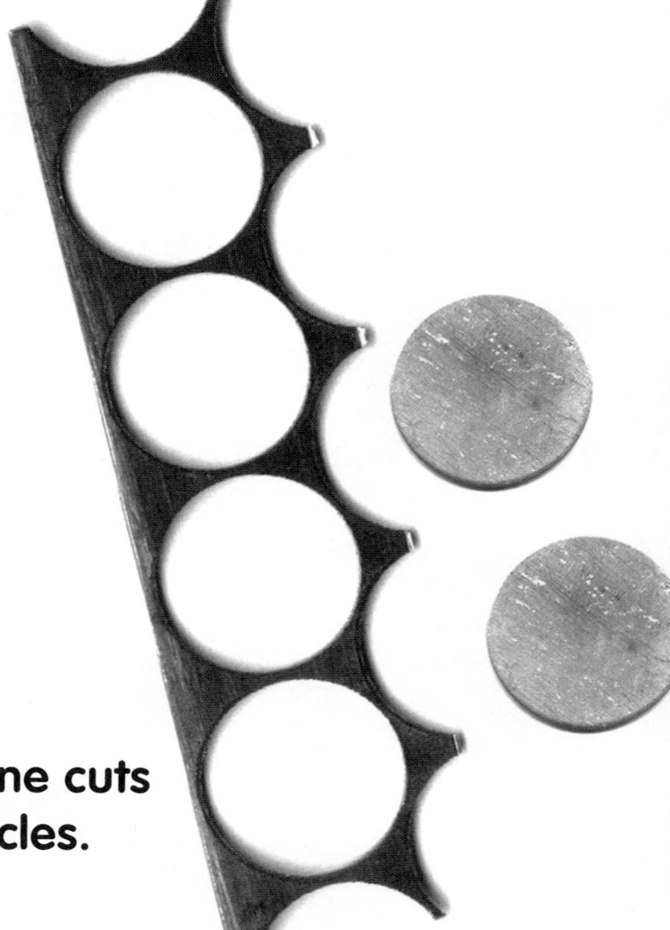

Then a machine cuts metal into circles.

▲ The circles are then stamped by a coin press.

▲ Lastly, they are placed into coin bags. Then they are sent to banks.

Each state has its own quarter.

Talk About It

Infer. How are coins made?

Tech Activity

Classroom Properties

What to Do

1. **Observe.** Look around your classroom. Collect five items.
2. **Record Data.** Make a chart like the one below. Draw the items you collected. Check off which properties each item has.

Draw Conclusions

Did any of your items have the same properties?

	eraser	marble	binder clip	paper clip	yarn
soft					✓
hard		✓	✓	✓	
smooth		✓	✓	✓	
rough	✓				
fuzzy					✓
round		✓			✓
square	✓				

Think, Talk, and Write

Complete the sentences below.

properties

natural resources

1. Materials made by nature that come from the Earth are called ________.

2. How something looks, feels, smells, and tastes are its ________.

3. What is cotton good for? What is it not good for?

4. Write about what you ate for breakfast. Where did the foods come from?

Art Link

Draw your favorite toy. What is it made of? If you do not know, ask an adult. Then write it under the picture.

LOG ON e-Tech Link For more on tools check out www.macmillanmh.com

Lesson 3

Vocabulary

system

electricity

Parts Work Together

Many objects have parts. Each part is important. The parts work together to make the object work. If just one part is missing or broken, the object will not work!

A wheel is one part of this object.

Which one of these trucks will work? How do you know?

A **system** is a group of parts that work together. A highway system has roads, streetlights, and cars. These parts work together. They get people where they want to go.

✔ **Can you think of some other systems?**

highway system

Systems at Work

A computer is a system, too. It has parts that work together.

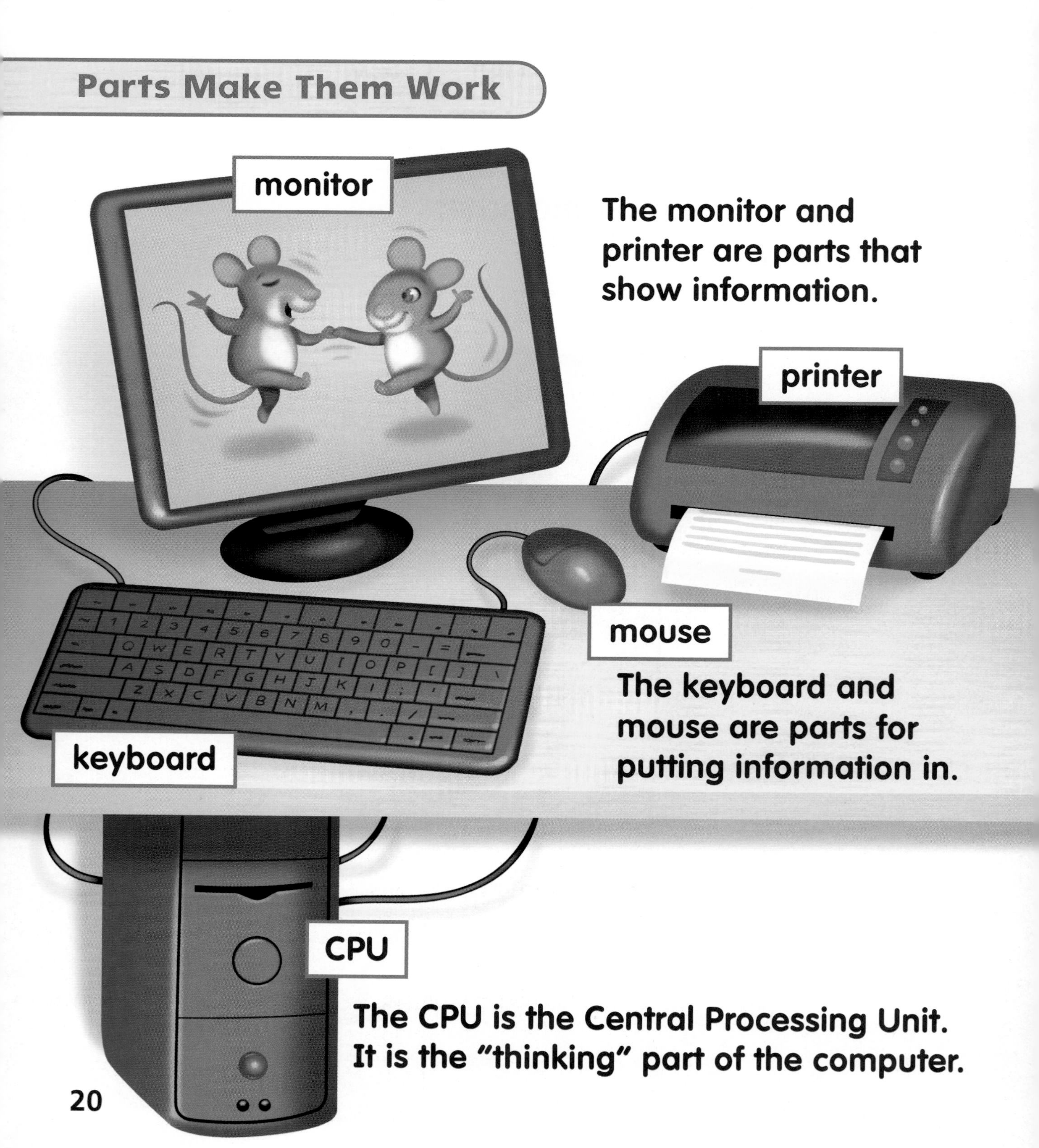

Parts Make Them Work
monitor
The monitor and printer are parts that show information.
printer
mouse
The keyboard and mouse are parts for putting information in.
keyboard
CPU
The CPU is the Central Processing Unit. It is the "thinking" part of the computer.

A lamp is also a system. It has parts that work together to make light.

A lamp has a lightbulb. It also has an electric cord that comes out of the the lamp. The cord plugs into an outlet in the wall. **Electricity** makes the lamp work. Now you have light!

light bulb

switch

✔ **What could be wrong if a lamp will not turn on?**

Read a Diagram

What would happen if a computer's keyboard was broken?

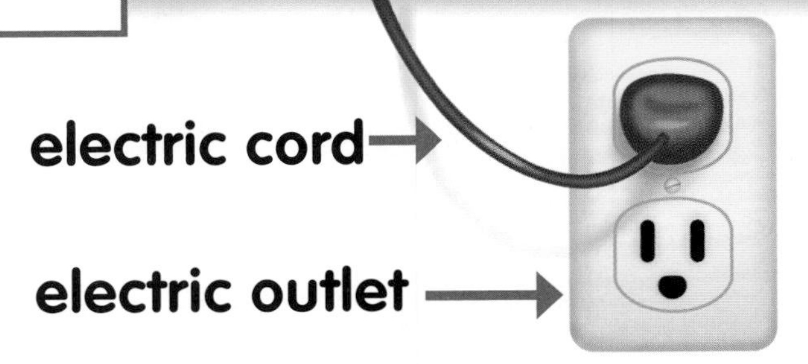

Put Parts Together

Wagons have parts.
When put together, all of these parts make a wagon.
On their own they are just pieces.

These are all of the parts of a wagon.

You can put an axle and wheel together. This lets the wheels spin.

Then, you can put the box above the wheels. This lets you put things in the wagon.

Lastly, you can attach a handle. This lets you pull the wagon.

Talk About It

Cause and Effect. What might happen if the wagon did not have a handle?

Tech Activity

You need

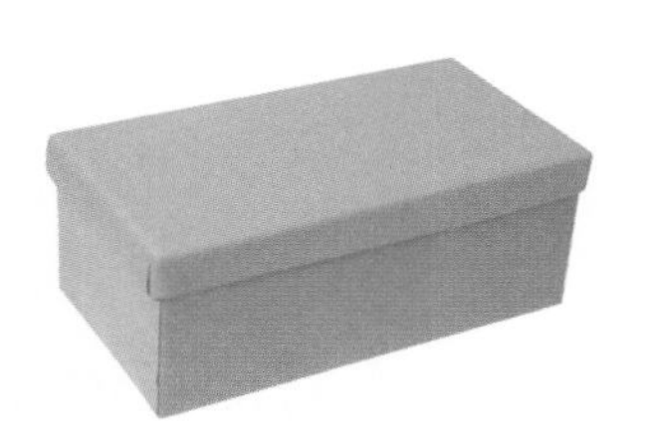

shoe box

coffee lids

paper fasteners

string

Make a Wagon

Make your own toy wagon!

What to Do

1. Get four circles from your teacher.
2. Hold each circle against the side of the box. Make sure the circles hang a little below the box.
3. Find the dot on the circle. △ **Be Careful.** Stick a paper fastener through the dot and the box.
4. △ **Be Careful.** Poke a hole at one end of the box. Use string to make a handle.
5. Decorate your wagon.

Draw Conclusions

How could you make your wagon work better?

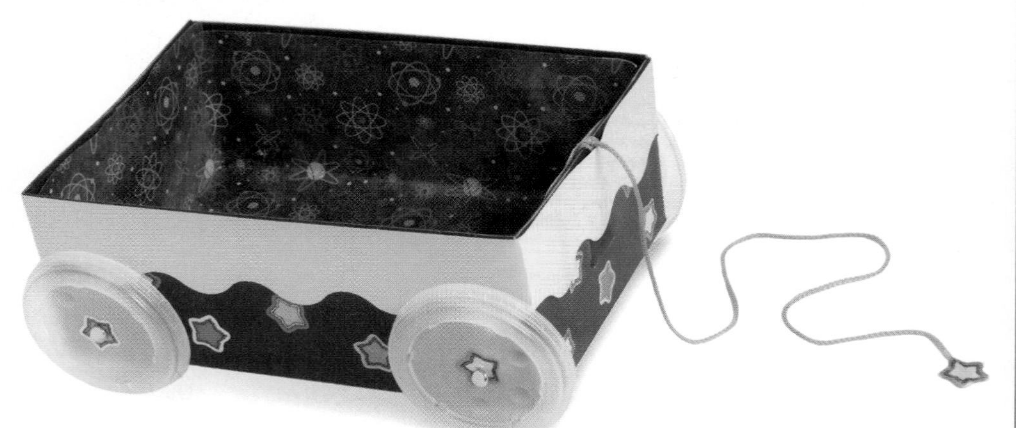

Think, Talk, and Write

Complete each sentence.

electricity
system

1. A flow of energy that makes a lamp work is ________.

2. A printer is a part of a ________ system.

3. Would this toy truck work? Why or why not?

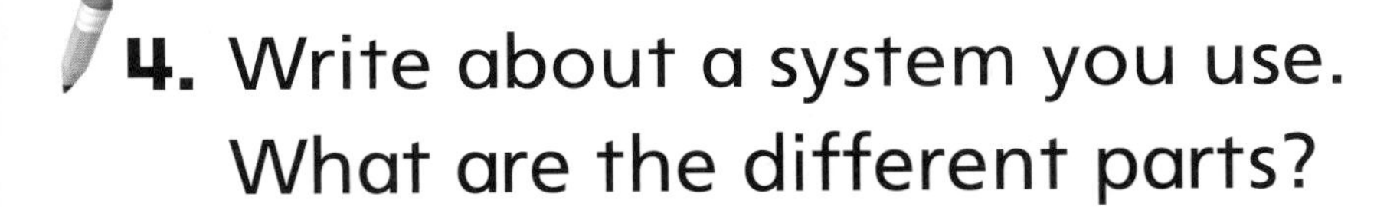

4. Write about a system you use. What are the different parts?

Art Link

Draw something you use every day that has parts. Try to name the parts. If you do not know what they're called, ask an adult for help.

LOG ON e-Tech Link For more on how parts work together check out www.macmillanmh.com

Lesson 4

Vocabulary

invent
inventor
communicate

Whose Idea Was That?

To **invent** something, you must start with an idea.

An **inventor** is a person who makes something for the very first time. Inventors see a need or problem in their everyday lives. Then they invent a way to solve it.

Alexander Graham Bell

This is the telephone Alexander Graham Bell invented.

The telephone made it easier for people to communicate. To **communicate** means to send and receive messages.

Telephones today do not look like the phone Bell invented. Over time, inventors made phones faster and easier to use.

✔ **Why was the telephone a good invention?**

Phones today let you communicate with people all over the world.

Italy

Improving Ideas

Anyone can invent things. It does not matter who you are or where you are from. People from all around the world have contributed to science and technology.

Improving Communication

Long ago people painted symbols on stones. This helped people communicate with each other.

The Egyptians invented a form of paper. This paper was easier to carry than stones.

Johannes Gutenberg improved the printing press. This made books available to more people.

Sometimes inventors improve things that already exist. This helps make people's lives easier.

Why was the typewriter an important invention?

Then, the typewriter was invented. This made writing faster and easier to read.

Today we use computers. Computers let people around the world communicate quickly.

Read a Diagram

What do you think will be the next invention?

Turn it On!

Thomas Edison and Lewis Latimer were inventors. They discovered how to make a better light bulb. Their light bulb could shine for a long time. It was also cheap enough for lots of people to buy.

Lewis Lattimer invented the long lasting filament inside the light bulb.

This is the light bulb Thomas Edison invented.

Today, there are many different types of light bulbs. People use different light bulbs for different amounts of light.

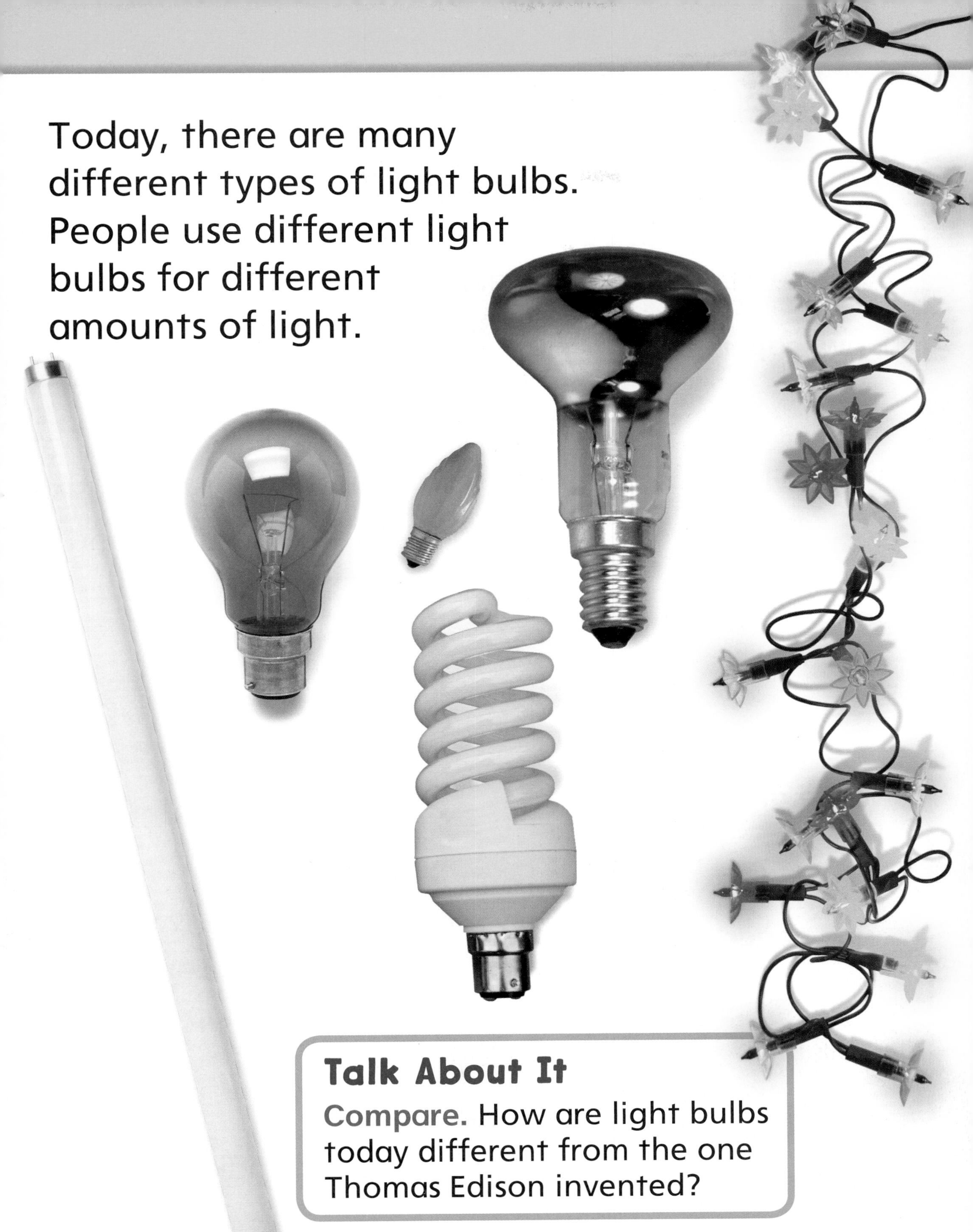

Talk About It

Compare. How are light bulbs today different from the one Thomas Edison invented?

Tech Activity

newspaper

bowl

water

white glue

wax paper

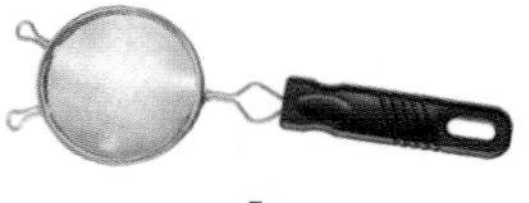
strainer

roller

Making Paper

What to Do

1. Tear very small pieces of newspaper into a bowl.
2. Add water until all the pieces are wet. Add a small squirt of glue.
3. Stir the mixture until it turns to mush.
4. Pour the mush into a strainer and squeeze the water out.

5. Spread the mush on to wax paper. Use a roller to flatten it. Lay another piece of wax paper on top. Cover it with a large book.

Draw Conclusions

Infer. Why is it good to make your own paper?

Think, Talk, and Write

Complete each sentence.

communicate
inventor

1. Telephones help people ________ with each other.

2. A person who makes something for the very first time is an ________.

3. Put these inventions in order.

4. Make up your own invention. Write about it.

Art Link

Make your own cave drawing.
Show how you get to school.

LOG ON e-Tech Link For more on inventros and invention check out www.macmillanmh.com

Lesson 5

Vocabulary

solution
design
model

From Idea to Invention

Aimee had a problem. She had to bring her pet hamster to the animal doctor. She needed something to carry him in.

Aimee needed to design a **solution**, or a way to fix her problem. To **design** is to draw, plan, build, and test an idea.

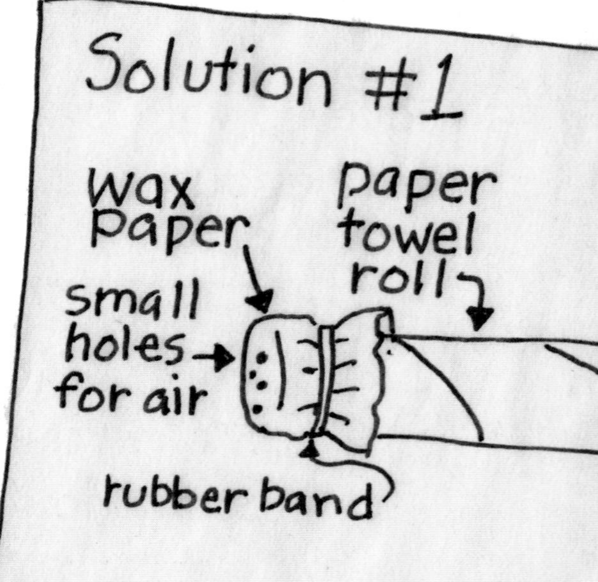

Aimee thought about different things she could use to carry her hamster. These ideas were possible solutions to her problem. She drew pictures of each of her ideas.

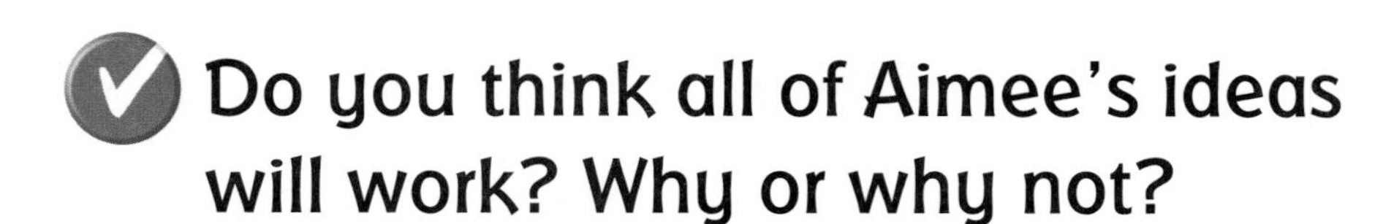

Do you think all of Aimee's ideas will work? Why or why not?

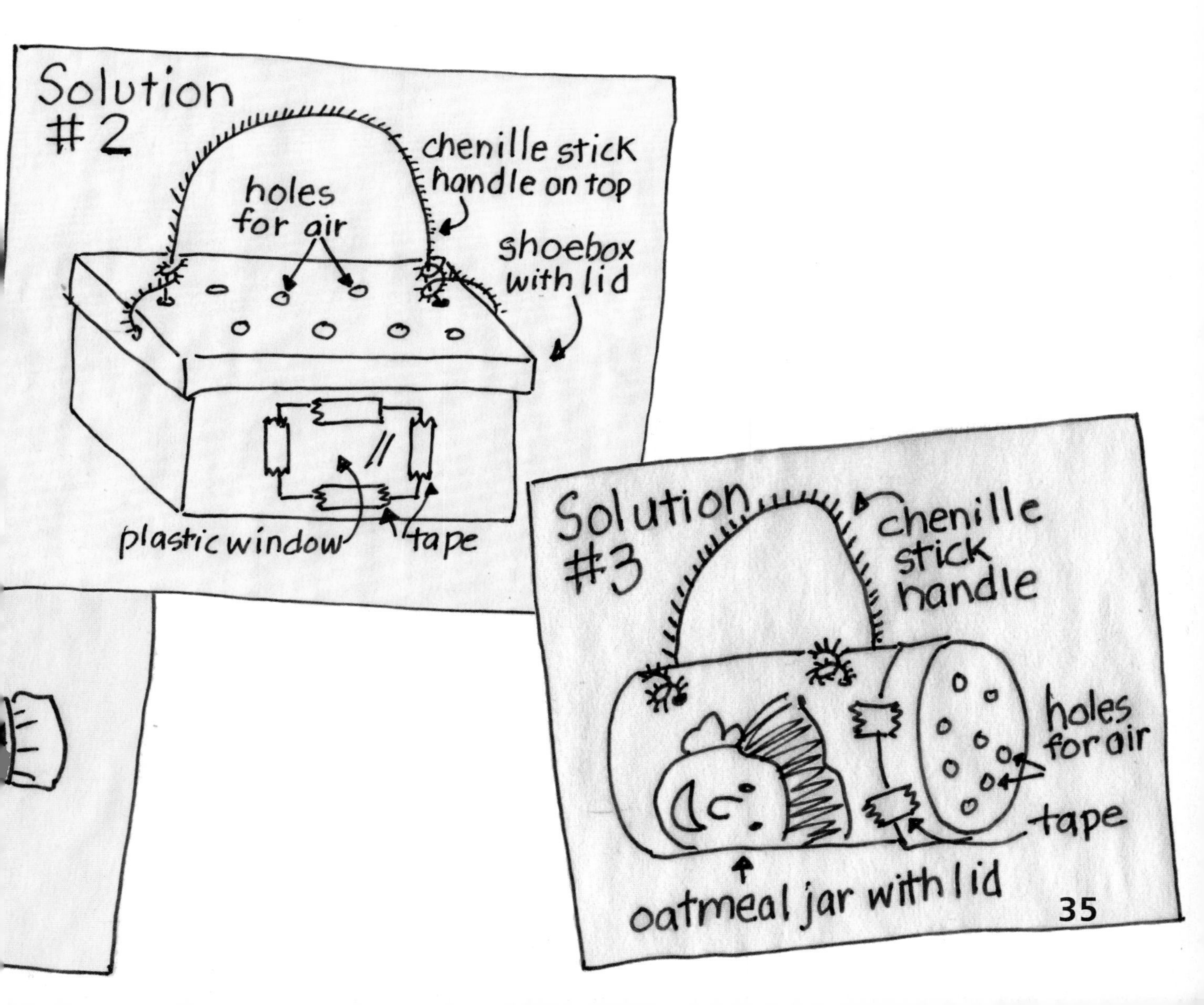

Making a Model

Aimee decided that the shoe box was the best idea for her hamster carrier. So, Aimee followed her drawing and made a **model**.

Do you see any problems with this carrier?

After Aimee made the shoe box, she realized she had to fix some things on it. Aimee moved the handles. She put them on the box instead of on the lid.

Now, the shoe box is a solution for Aimee to carry her hamster to the doctor.

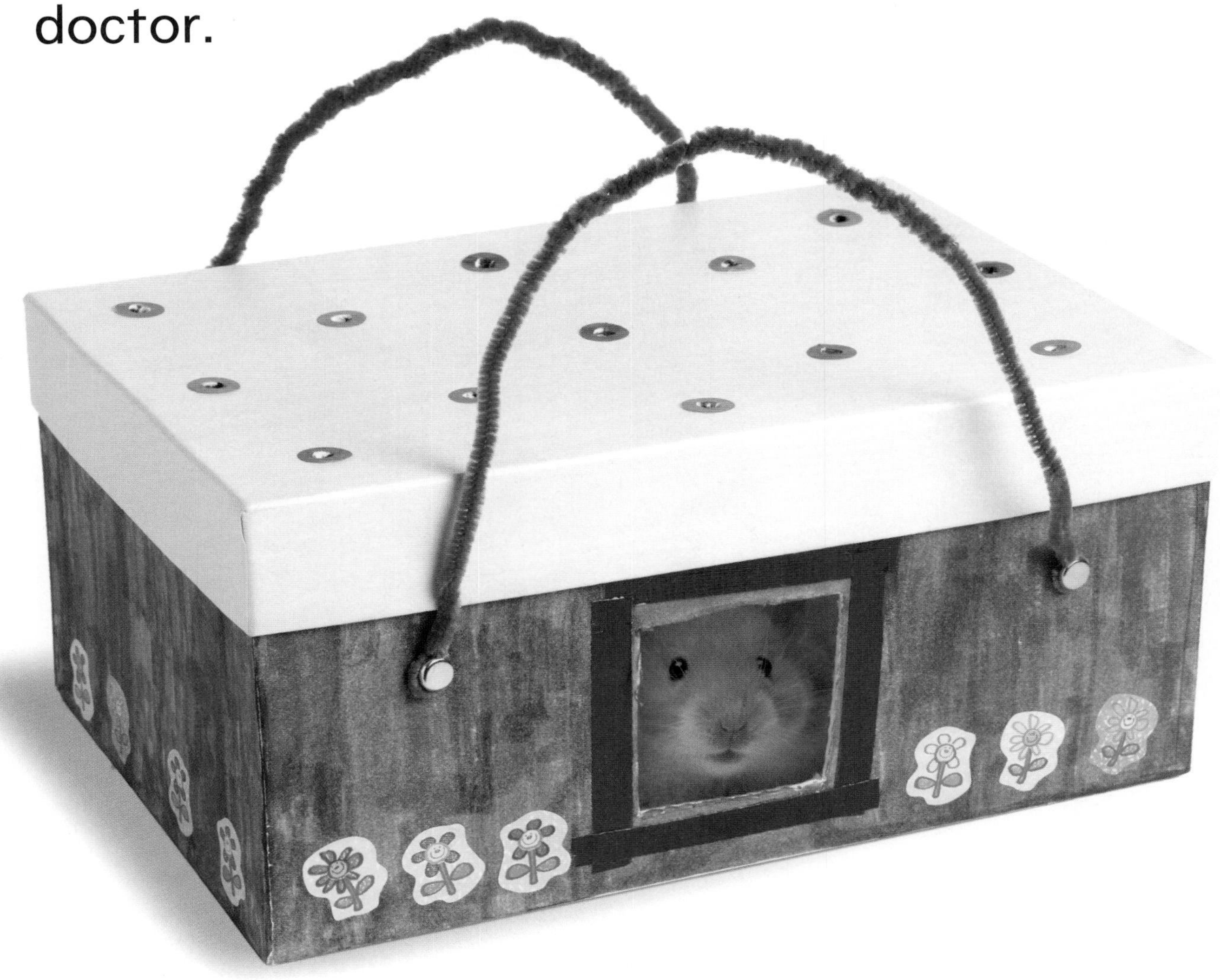

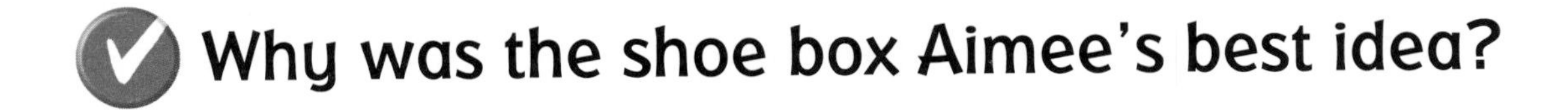

Why was the shoe box Aimee's best idea?

Testing, Testing

When inventors make something, they have to test it. They test it to make sure it is perfect. Sometimes, inventors have to make something many times before they get it right.

Sabrina the inventor had to make lots of toys before she figured out which one was the best.

All new toy ideas have to be tested. Inventors must prove that their toys are safe. The parts can not be too small.

Sometimes children test toys to make sure they are fun and safe to play with.

Now the toys are in stores. Children can buy the toys and enjoy them!

Talk About It

Infer. How would you test a new kind of toy truck?

Tech Activity

You need

newspaper

aluminum foil

bubble wrap

ice cubes

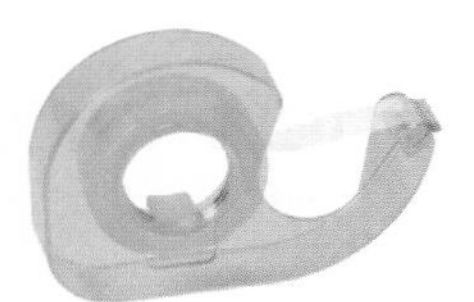

tape

Design a Juice Box!

What to Do

1. Wrap three ice cubes in three different materials. Wrap one ice cube in newspaper, one in aluminum foil, and one in bubble wrap. Use the same amount of material each time.

2. **Observe.** Wait one hour. Unwrap each ice cube. What do you notice?

3. **Communicate.** Describe what happened to each ice cube. Which ice cube melted the most? Which one melted the least? Why?

Draw a Conclusion

Which material would be the best for keeping juice cool? Why?

Think, Talk, and Write

Complete each sentence.

solution
design

1. A way to fix a problem is a ________.

2. To ________ an idea is to draw, plan, and build it.

3. Write about how Aimee figured out the best way to carry her hamster to the doctor.

4. Why would this carrier not work?

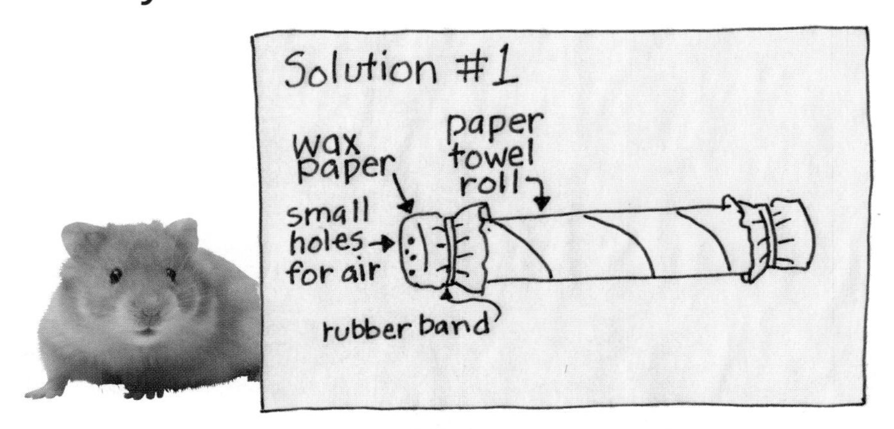

Art Link

Draw an idea for a new toy or a change to an old one. What problem are you trying to fix?

Do You Remember?

Where else in this book did you see each picture? Tell what you read about it.

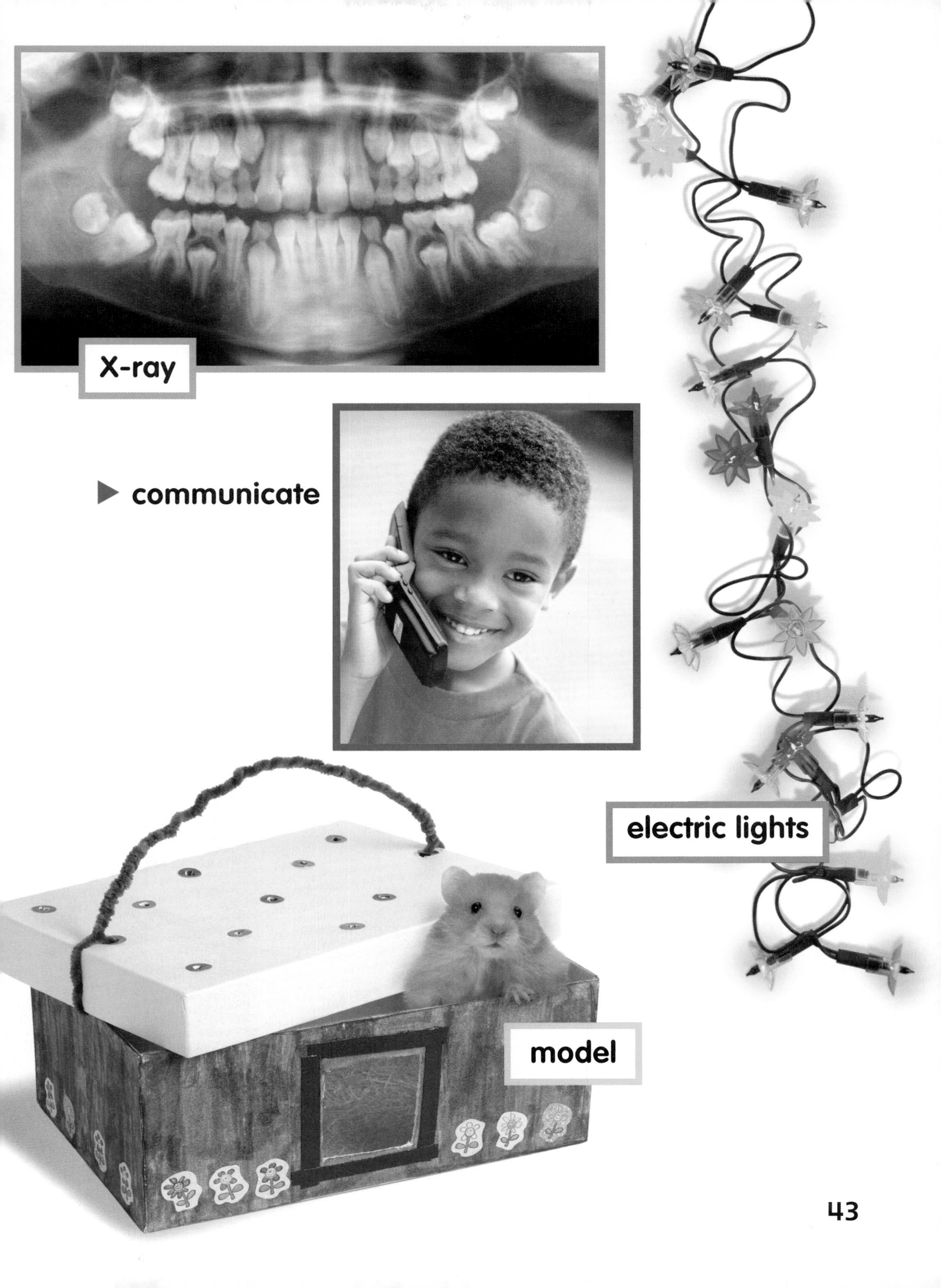
X-ray
▶ communicate
electric lights
model

Glossary

C

communicate To send and receive messages. (p. 27) **People communicate using technology.**

D

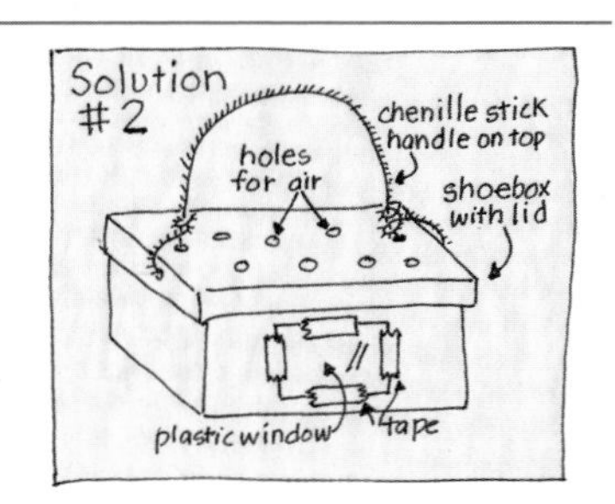

design To draw, plan, build, or test an idea. (p. 34) **Aimee designed a box to hold her hamster.**

E

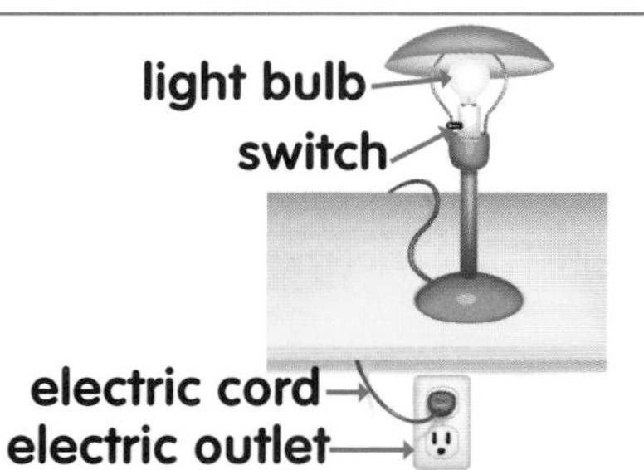

electricity A flow of energy that makes things work. (p. 21) **Electricity flows from outlets to appliances like lamps.**

I

invent To think of something, or make it for the first time. (p. 26)

inventor A person who invents, or makes something for the first time. (p. 26)

M

mint A place where money is made. (p. 14)

model A sample of a product or idea, used for testing. (p. 36)

N

natural resources Materials from nature that people use. (p. 10)

P

property How something looks, feels, smells, or tastes. (p. 12) **Softness is a property of cotton.**

S

solution A way to fix a problem. (p. 34) **Aimee's pet carrier was a solution to her problem.**

system A group of parts that work together. (p. 19) **Cars, roads, and street lights are parts of the highway system.**

T

technology All the tools we use to make life better. (p. 3)

tools An object that helps you do work. (p. 2)

X

X-ray A light wave that can help doctors see inside the body. (p. 6)

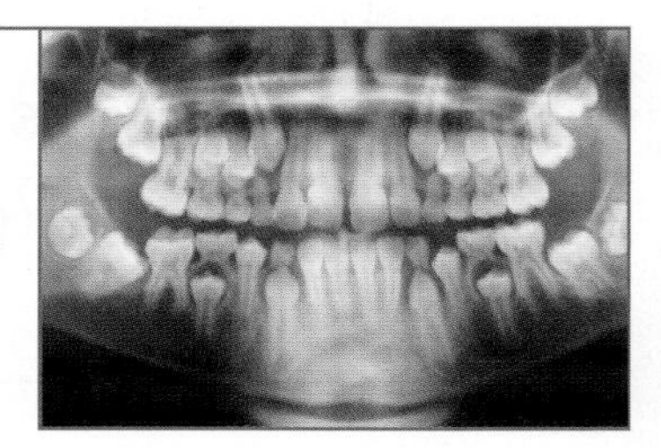

Credits

Illustration Credits: 12, 20-21, 23, 38, 39, 44: David Clegg.

All Photography Credits: All photographs are by Macmillan/McGraw-Hill (MMH) except as noted below:

ii: (t) United States Mint; (tcr) United States Mint; (tcl) United States Mint; (tr) United States Mint; (c) Ariel Skelley/Getty Images; (b) SSPL / The Image Works. iii: (c) Mark Boulton / Alamy; (t) Stephen Ogilvy for MMH; (br) G.K. & Vikki Hart/Getty Images. 2: (l) Mike Watson Images/Punchstock; (r) The Stock Asylum, LLC/Alamy; (b) Stephen Ogilvy for MMH. 3: (bl) Tom Bean/CORBIS; (br) Photo by Gene Alexander, USDA Natural Resources Conservation Service; (t) C Squared Studios/Getty Images. 4: (t) Burke/Triolo/BrandXPictures/ Jupiter Images; (b) Gabe Palmer/CORBIS. 5: (t) Emma Lee/Life File/Getty Images; (b) Erin Patrice O'Brien/Getty Images; (c) Siede Preis/Getty Images. 6: (l) Roy Morsch/CORBIS; (r) Custom Medical Stock Photo; (t) Mediscan/CORBIS. 7: (l) Mediscan/CORBIS; (tr) Michael Newman/PhotoEdit. 8: Myrleen Ferguson Cate/PhotoEdit. 9: (l) Mediscan/CORBIS; (r) C Squared Studios/Getty Images. 10: (tl) Siede Preis/Getty Images; (bl) C Squared Studios/Getty Images; (tr) Royalty-Free/CORBIS; (br) Photodisc/Punchstock. 11: (tl) Scientifica / Visuals Unlimited; (bl) C Squared Studios/Getty Images; (tc) Siede Preis/Getty Images; (bc) Corbis/Punchstock; (tr) Eureka/Alamy; (br) David Toase/Getty Images. 12: (t) The McGraw-Hill Companies, Inc./Ken Karp photographer; (tcl) Royalty-Free/CORBIS; (tc) Thomas Northcut/Photodisc/Getty Images; (tcr) Stockdisc/Getty Images. 13: (tc) Bill Aron/PhotoEdit; (tcl) Stockbyte/PictureQuest; (t) Brand X Pictures/Punchstock; (tcr) The McGraw-Hill Companies Inc./Ken Cavanagh, Photographer; (b) Ariel Skelley/Getty Images; (tl) Stockdisc/ Punchstock; (tr) Photodisc/Getty Images. 14: (c) MMH; (l) Gianni Cigolini/Getty Images; (r) US Mint. 15: (tl) Justin Sullivan/Getty Images; (bl) United States Mint; (tr) James L. Amos/CORBIS; (r) US Mint; (bcl) United States Mint; (bcl) United States Mint; (bcl) United States Mint. 16: (bcl) Dorling Kindersley/Getty Images; (t) The McGraw-Hill Companies Inc./Ken Cavanagh, Photographer; (bc) Royalty-Free/ CORBIS; (tc) Ryan McVay/Photodisc/Getty Images; (tr) Ingram Publishing/Alamy. 17: (t) David Toase/Getty Images; (c) The McGraw-Hill Companies, Inc./Ken Karp photographer; (b) Photodisc/Punchstock. 19: Scott Tysick/Masterfile. 22: GK & Vikki Hart/Getty Images. 23: (inset) G.K. & Vikki Hart/Getty Images; (b) GK & Vikki Hart/Getty Images. 26: (l) The Granger Collection, New York; (r) SSPL/The Image Works. 27: (cl) George Shelley/CORBIS; (cr) Rolf Bruderer/Getty Images; (b) WorldSat International Inc./Photo Researchers, Inc.; (t) Brand X Pictures/Punchstock. 28: (l) Neil Beer/Getty Images. 28: (c) Bildarchiv Preussischer Kulturbesitz/Art Resource, NY; (r) Erich Lessing / Art Resource, NY. 29: (l) Burke/Triolo/BrandXPictures/Jupiter Images; (r) Siede Preis/Photodisc/PictureQuest. 30: (bl) Bettmann/ CORBIS; (t) Photri/Microstock; (br) Bettmann/CORBIS. 31: (bc) Mark Boulton / Alamy; (l) Jerry Young/Dorling Kindersley; (tcl) Peter Anderson/Dorling Kindersley; (tc) C Squared Studios/Getty Images; (tcr) 42pix/Alamy; (r) Steve Shott and Trish Gant/Dorling Kindersley. 32: Burke/Triolo Productions/Getty Images. 33: (l) Burke/Triolo/BrandXPictures/Jupiter Images; (c) Siede Preis/Photodisc/PictureQuest. (r) Neil Beer/Getty Images. 34: PhotoSpin, Inc/Alamy. 36: (inset) BIOS Klein & Hubert/Peter Arnold, Inc. 37: (inset) G.K. & Vikki Hart/ Getty Images. 41: G.K. & Vikki Hart/Getty Images. 42: (tl) Scott Tysick/Masterfile; (bl) Eureka/Alamy; (tr) Burke/Triolo/BrandXPictures/ Jupiter Images; (br) Bettmann/CORBIS. 43: (tl) Mediscan/CORBIS; (cl) George Shelley/CORBIS; (r) Steve Shott and Trish Gant/Dorling Kindersley. 44: (1) George Shelley/CORBIS; (4) SSPL/The Image Works; (5) .; (6) James L. Amos/CORBIS. 45: (2) Royalty-Free/CORBIS; (3) The McGraw-Hill Companies, Inc./Ken Karp photographer; (1) BIOS Klein & Hubert/Peter Arnold, Inc.; (4) G.K. & Vikki Hart/Getty Images; (5) Scott Tysick/Masterfile; (6) Mike Watson Images/Punchstock. 46: (2) Mediscan/CORBIS; (1) The Stock Asylum, LLC/Alamy.